Fieldwork and Dissertations in Geography

Geography Discipline Network (GDN)

Higher Education Funding Council for England
Fund for the Development of Teaching and Learning

Dissemination of Good Teaching, Learning and Assessment Practices in Geography

Aims and Outputs

The project's aim has been to identify and disseminate good practice in the teaching, learning and assessment of geography at undergraduate and taught postgraduate levels in higher education institutions.

Ten guides have been produced covering a range of methods of delivering and assessing teaching and learning:

- Teaching and Learning Issues and Managing Educational Change in Geography
- Lecturing in Geography
- Small-group Teaching in Geography
- Practicals and Laboratory Work in Geography
- Fieldwork and Dissertations in Geography
- Resource-based Learning in Geography
- Teaching and Learning Geography with Information & Communication Technologies
- Transferable Skills and Work-based Learning in Geography
- Assessment in Geography
- Curriculum Design in Geography

A resource database of effective teaching, learning and assessment practice is available on the World Wide Web, http://www.chelt.ac.uk/gdn, which contains national and international contributions. Further examples of effective practice are invited; details regarding the format of contributions are available on the Web pages. Examples should be sent to the Project Director.

Project Team

Lead site: ***Cheltenham & Gloucester College of Higher Education***
Professor Mick Healey; Dr Phil Gravestock; Dr Jacky Birnie; Dr Kris Mason O'Connor

Consortium: ***Lancaster University***
Dr Gordon Clark; Terry Wareham
Middlesex University
Ifan Shepherd; Professor Peter Newby
Nene — University College Northampton
Dr Ian Livingstone; Professor Hugh Matthews; Andrew Castley
Oxford Brookes University
Dr Judy Chance; Professor Alan Jenkins
Roehampton Institute London
Professor Vince Gardiner; Vaneeta D'Andrea; Shân Wareing
University College London
Dr Clive Agnew; Professor Lewis Elton
University of Manchester
Professor Michael Bradford; Catherine O'Connell
University of Plymouth
Dr Brian Chalkley; June Harwood

Advisors: Professor Graham Gibbs (*Open University, Milton Keynes*)
Professor Susan Hanson (*Clark University, USA*)
Dr Iain Hay (*Flinders University, Australia*)
Geoff Robinson (*CTI Centre for Geography, Geology and Meteorology, Leicester*)
Professor David Unwin (*Birkbeck College, London*)
Dr John Wakeford (*Lancaster University*)

Further Information

Professor Mick Healey, Project Director   Tel: +44 (0)1242 543364   Email: mhealey@chelt.ac.uk
Dr Phil Gravestock, Project Officer   Tel: +44 (0)1242 543368   Email: pgstock@chelt.ac.uk
Cheltenham & Gloucester College of Higher Education
Francis Close Hall, Swindon Road, Cheltenham, GL50 4AZ, UK   [Fax: +44 (0)1242 532997]

http://www.chelt.ac.uk/gdn

Fieldwork and Dissertations in Geography

Ian Livingstone, Hugh Matthews and Andrew Castley

Nene — University College Northampton

Series edited by Phil Gravestock and Mick Healey
Cheltenham & Gloucester College of Higher Education

Published by:

Geography Discipline Network (GDN)
Cheltenham & Gloucester College of Higher Education
Francis Close Hall
Swindon Road
Cheltenham
Gloucestershire, UK
GL50 4AZ

Fieldwork and Dissertations in Geography

ISBN: 1 86174 028 X ✓
ISSN: 1 86174 023 9

Typeset by Phil Gravestock

Cover design by Kathryn Sharp

Printed by:

Frontier Print and Design Ltd.
Pickwick House
Chosen View Road
Cheltenham
Gloucestershire, UK

Contents

Editors' preface

This Guide is one of a series of ten produced by the Geography Discipline Network (GDN) as part of a Higher Education Funding Council for England (HEFCE) and Department of Education for Northern Ireland (DENI) Fund for the Development of Teaching and Learning (FDTL) project. The aim of the project is to disseminate good teaching, learning and assessment practices in geography at undergraduate and taught postgraduate levels in higher education institutions.

The Guides have been written primarily for lecturers and instructors of geography and related disciplines in higher education and for educational developers who work with staff and faculty in these disciplines. For a list of the other titles in this series see the information at the beginning of this Guide. Most of the issues discussed are also relevant for teachers in further education and sixth-form colleges in the UK and upper level high school teachers in other countries. A workshop has been designed to go with each of the Guides, except for the first one which provides an overview of the main teaching and learning issues facing geographers and ways of managing educational change. For details of the workshops please contact one of us. The Guides have been designed to be used independently of the workshops.

The GDN Team for this project consists of a group of geography specialists and educational developers from nine old and new universities and colleges (see list at front of Guide). Each Guide has been written by one of the institutional teams, usually consisting of a geographer and an educational developer. The teams planned the outline content of the Guides and these were discussed in two workshops. It was agreed that each Guide would contain an overview of good practice for the particular application, case studies including contact names and addresses, and a guide to references and resources. Moreover it was agreed that they would be written in a user-friendly style and structured so that busy lecturers could dip into them to find information and examples relevant to their needs. Within these guidelines the authors were deliberately given the freedom to develop their Guides in their own way. Each of the Guides was refereed by at least four people, including members of the Advisory Panel.

The enthusiasm of some of the authors meant that some Guides developed a life of their own and the final versions were longer than was first planned. Our view is that the material is of a high quality and that the Guides are improved by the additional content. So we saw no point in asking the authors to make major cuts for the sake of uniformity. Equally it is important that the authors of the other Guides are not criticised for keeping within the original recommended length!

Although the project's focus is primarily about disseminating good practice within the UK a deliberate attempt has been made to include examples from other countries, particularly North America and Australasia, and to write the Guides in a way which is relevant to geography staff and faculty in other countries. Some terms in common use in the UK may not be immediately apparent in other countries. For example, in North America for 'lecturer' read 'instructor' or 'professor'; for 'staff' or 'tutor' read 'faculty'; for 'postgraduate' read 'graduate'; and for 'Head of Department' read 'Department Chair'. A 'dissertation' in the

UK refers to a final year undergraduate piece of independent research work, often thought of as the most significant piece of work the students undertake; we use 'thesis' for the Masters/ PhD level piece of work rather than 'dissertation' which is used in North America.

In addition to the Guides and workshops a database of good practice has been established on the World Wide Web (http://www.chelt.ac.uk/gdn). This is a developing international resource to which you are invited to contribute your own examples of interesting teaching, learning and assessment practices which are potentially transferable to other institutions. The resource database has been selected for *The Scout Report for Social Sciences*, which is funded by the National Science Foundation in the United States, and aims to identify only the best Internet resources in the world. The project's Web pages also provide an index and abstracts for the *Journal of Geography in Higher Education*. The full text of several geography educational papers and books are also included.

Running a consortium project involves a large number of people. We would particularly like to thank our many colleagues who provided details of their teaching, learning and assessment practices, many of which appear in the Guides or on the GDN database. We would also like to thank, the Project Advisers, the FDTL Co-ordinators and HEFCE FDTL staff, the leaders of the other FDTL projects, and the staff at Cheltenham and Gloucester College of Higher Education for all their help and advice. We gratefully acknowledge the support of the Conference of Heads of Geography Departments in Higher Education Institutions, the Royal Geographical Society (with the Institute of British Geographers), the Higher Education Study Group and the *Journal of Geography in Higher Education*. Finally we would like to thank the other members of the Project Team, without them this project would not have been possible. Working with them on this project has been one of the highlights of our professional careers.

Phil Gravestock and Mick Healey

Cheltenham

July 1998

All World Wide Web links quoted in this Guide were checked in July 1998. An up-to-date set of hyperlinks is available on the Geography Discipline Network Web pages at:

http://www.chelt.ac.uk/gdn

About the authors

Ian Livingstone

I have worked as a lecturer in higher education since 1987 and am currently Reader in Earth Science at Nene — University College Northampton. I am joint editor of the *Journal of Geography in Higher Education* and run GeogNet, the email discussion list for UK geographers which is primarily concerned with issues of teaching and learning in he geography. I have been involved in a number of PCFC, HEFCE and DfEE initiatives on teaching and learning. My research interest is in aeolian geomorphology, particularly in deserts, and this work has taken me to southern and north Africa, Australia and Arabia.

Hugh Matthews

I am Professor of Geography and Director of the Centre for Children and Youth at Nene — University College Northampton. For the last three years I have been joint editor of the *Journal of Geography in Higher Education.* My interests in teaching, learning and assessment in geography arose out of my courses in human geography at Coventry University, where I worked for 19 years. During this time I was part of a team which was awarded the British Petroleum Prize for teaching innovation in geography. Since 1992 I have been Secretary of the Higher Education Study Group of the RGS-IBG and I acted as a HEFCE Assessor in the Teaching Quality Assessment of Geography. Between 1994-1996 I was Director of the Geography Discipline Network, an initiative funded by the DfEE. This network considered notions of student empowerment, life-long learning and employer links in geography. Apart from my interests in educational development, my research focuses on the geography of children and young people's rights to participation. Currently, I am working on the ESRC programme, 'Children 5-16: growing into the 21st Century'.

Andrew Castley

After graduating in German and French in 1969, I went into industrial exporting for 4 years before switching to teaching languages for business purposes for 15 years. From 1988 to 1995 I led the BA/BSc combined honours course at Nene — University College Northampton then moved into staff and educational development. I am currently the director of the office for education and staff development at Nene — University College Northampton. I have published text books in languages for business and on peer learning.

Introduction to this Guide

This Guide is in two units: one deals with fieldwork and the other with dissertations. These units stand apart from each other and can be treated as separate Guides. However, both fieldwork and dissertations have often been viewed as central elements of geography degree courses in the UK, and both have been the subject of considerable scrutiny and reappraisal in the recent past. The purpose of this Guide is to question and explore the nature, purpose and relevance of fieldwork and dissertations as components of a geography degree. In both cases there is no model or prescriptive practice. Indeed, the evidence of the Teaching Quality Assessment (TQA) is that diversity and variety are features of the way in which the geography curricula are presented within UK higher education institutions. This is not to say that all fieldwork and all dissertation work is implicitly 'good'. There is a danger that because both are traditional features of most degree programmes they become elements which are taken for granted and in so doing, encourage uncritical thought. Some of the issues we will focus on are introduced with the aim of getting geography providers to think about why, how and when fieldwork and dissertations should be undertaken, if at all. Given the changing nature of higher education, especially in relation to higher student:staff ratios, and the growing clamour that curricula should be more keenly honed on the skills required by employers, it seems apposite to reflect about the delivery and form of these components. Because of these sentiments, this Guide is designed to enable the sharing of current good practice. All the examples have been taken from existing curricula and we have relied on case studies which demonstrate that there are many ways of achieving common goals. What we would like to think is that after having read this Guide you will be better-placed to compare how your course relates to what is going on elsewhere and, if you are tempted to change some of what you are doing, it is because the Guide has made you think about some aspect of your course in a slightly different way.

Unit 1: Fieldwork

1 Introduction to this Unit

1.1   Why do fieldwork?

Fieldwork is essentially the opportunity for students to gain first-hand experience of geographical issues away from the classroom. In this Guide we use the term to cover the full range of activities which extend from a short trip out from the institution, possibly during timetabled class time, through half-day and day trips, to larger scale fieldcourses. These include regular trips out throughout the duration of a course element or module and the traditional, residential fieldcourse spread over several days, a week or even longer. Some fieldwork involves close staff guidance and supervision, while on other occasions students may be devising and executing projects largely independently of staff.

Whatever the pattern of delivery, fieldwork is often seen by geographers as an essential element within the geography curriculum, but it can be viewed as little more than an excuse to avoid 'proper' teaching or even as an expenses-paid holiday by many resource managers. It may therefore be useful to consider why we do fieldwork. This is not only so that we can justify the budget, but also so that we become more focused about the purpose of our fieldtrips.

In the past geographers seldom articulated why they did fieldwork, although lots of us were sure it was a good thing. Many of us greatly enjoy field teaching and often it is the first thing that graduate geographers can recall about their course. However, (perhaps sadly) we cannot justify it simply on the grounds that it was memorable or because we and the students enjoyed it.

Reasons given for undertaking fieldwork

Here is a list of reasons for doing fieldwork (an adaptation and extension of a list from Gold *et al.*, 1991, pp.25-26):

- develops observational skills;

- allows students to visit places they would not normally see;

- facilitates experiential learning through a focus on the real world — comparing real examples with model or idealised examples in textbooks;

- encourages students to take responsibility for their own learning;

- develops analytical skills;

- experiencing 'real' research through fieldwork, particularly residential fieldwork, allows concentrated study of a topic not usually possible with the normal timetable structure;

- develops a respect for the environment;

- develops personal skills, such as teamwork, leadership and responsibility;

- breaks down barriers — fieldwork often facilitates more relaxed social contact between students and their peers and between students and staff.

Think about these reasons. Are there others which ought to be in the list? Can you rank their importance to you? Is any single reason sufficient to justify fieldwork? Think of your ranked list as course objectives (or learning outcomes). If the reasons are viewed as objectives, are there other ways these objectives could be fulfilled? Is any one of these only achievable through fieldwork? How would geography students be poorer for not doing fieldwork?

1.2   What is the aim of this unit?

Assuming that we believe that fieldwork is worth doing, it should be possible for us to learn from others. A glance at recent issues of the *Journal of Geography in Higher Education* or the *Professional Geographer* shows that fieldwork is a topic of recurrent interest to contributors to these journals, and that there is much to debate. As with other aspects of our courses, recent changes in the resourcing and delivery of higher education mean that we have had to re-evaluate the fieldwork elements of our courses. The aim of this unit of the Guide is to open discussion on some of the key issues in fieldwork provision in higher education geography courses. Section 2 of this unit deals with these issues. Some of the issues are resource-based and some are a response to more general pedagogic innovations in the subject and more widely in the sector. Section 3 provides some case studies and Section 4 suggests some sources of further information.

2   Issues for successful fieldwork

For all higher education geographers who are concerned about their teaching the central issue is how to improve the quality of their students' learning. Some of the ways of improving teaching and learning involve organisational, administrative or structural changes at a national, institutional or sub-institutional level, but others are within the control (to a greater or lesser extent) of the teachers and students. The material covered in this unit on issues for successful fieldwork covers some of the areas which individual teachers might consider and in which they can reasonably be expected to have an impact.

2.1   Purpose of fieldwork

In Section 1.1 on "Why do fieldwork?" we started to consider the justification for doing fieldwork. This discussion can be developed to consider where fieldwork fits in the curriculum. In the past it was often felt that getting students out and about was sufficient. Increasingly it is apparent that good curriculum design requires either that the fieldtrip fits alongside some other part of the curriculum to complement, extend or enhance it, or that the course fulfils some major objective of the curriculum as a whole.

Consider the following questions (which relate to wider questions of curriculum design — see the Guide in this series on "Curriculum Design in Geography"):

- Is this fieldtrip essentially stand-alone, in that it is the only element of the course which fulfils a major stated objective of the course?

- Do other parts of the course also help to fulfil the same objectives?

- How does the fieldtrip build on students' previous knowledge and experience in the course?

- What do students need to learn (methods, techniques, factual information) which they will use later in the course?

- How complementary is fieldwork in different elements (modules) of the course or at different levels (years)? Is there a 'spiral curriculum' in which similar themes, issues, skills or techniques are revisited with increasing degrees of sophistication at subsequent levels of the course?

- How long should the fieldcourse last? Is there a need to get acquainted with the site before meaningful work can be executed? Do 'raids' on sites lead to good learning?

- Do our students view these experiences as outside the rest of the curriculum and fail to relate them to the geography course? Should we get students used to fieldwork early in the course so that the 'event' does not overshadow the learning?

2.2   Preparing students and following up fieldwork

Preparation

Some fieldwork is stand-alone so that all work is undertaken during a fieldcourse; other fieldcourses require students to prepare for the course in advance; some are designed to include follow-up work.

Whether the preparation is explicit or not, the student will be building on previous elements of the course (except, perhaps, in the case of introductory courses in the first or induction week: these fieldcourses have very explicit objectives). Fieldcourses rarely should, or do, stand in complete isolation from other elements of the course.

Consider the following questions. How have our students been prepared for this fieldcourse? What in their previous experience will be of explicit value on this trip? What more is required to ensure that they are enabled to undertake the tasks we are setting or fulfil the fieldcourse's objectives?

Some of the more direct ways of preparing students for fieldwork have included setting reading, giving lectures, and running workshops, seminars or tutorials. Warburton & Higgitt (1997) suggested ways in which information technology (IT) might be used in preparation for fieldwork. Their list includes: providing theoretical background information; providing site-specific information; outlining aims and objectives of the field activity; technical preparation; and logistical information. Using IT allows preparation to be self-paced and more student-centred, but does not automatically enhance the learning experience. In more detail they discuss producing a proto-type geographical information system (GIS) as a database for the field area and briefing tutorial material.

Follow-up sessions

Older-style fieldwork centred on some form of data collection, often to confirm some general model or hypothesis learnt elsewhere in the course. However, in addition it is often important to allow analysis, discussion and reflection. Traditionally, students have often been asked to write reports on their return from the fieldcourse. The choice is between self-contained fieldcourses where student output is entirely (or almost entirely) produced within the time-span of the fieldcourse and those requiring later assigned work and follow-up sessions. The pros and cons of these alternatives are considered in Section 2.7.

The fieldwork can also act as an integral part of a wider course or can be used to introduce themes and issues to be picked up later in the curriculum.

2.3   Types and styles of fieldwork

There are many different types and styles of fieldwork. Kent *et al.* (1997) recently provided a review of the changes in types and styles of fieldwork and its delivery to students. In general there is a move away from telling students about places to getting them to think through issues and implications. The range of types and styles can be distinguished by criteria such as: status in the undergraduate geography programme; nature of the skills training; teaching and learning methods (themes); assessment strategies; management practice; and resource provision (Table 1.1).

Table 1.1: *A classification of types and styles of fieldwork*

Status in the course	optional freestanding	compulsory integrated
Pattern of delivery	residential local	non-residential non-local
Skills training	implicit extensive passive course specific knowledge-based introductory	explicit focused active transferable process-based advanced
Delivery	staff-selected staff-led reception prescribed issue-based	student-negotiated student-led discovery opportunities for initiative problem-based
Assessment	traditional individual tutor-assessed directed self-contained	innovative group peer-assessed negotiated on-going
Management	controlled rigid whole class	creative flexible individual student
Resources	institution-financed unconstrained external links	student-financed constrained no external links

Adapted from McEwen (1996)

Consider how each of your fieldcourses fits into the classification in Table 1.1. Does your fieldwork provision change with each level (year) of the course? To what extent are these differences the outcomes of a teaching and learning strategy rather than an unplanned response to other constraints? Are there any changes you would like to see made to the types and styles of fieldwork?

Now fill in the following table (or your own version of it)

List of fieldcourses	Location	Duration	Style (use the list in Table 1.1 to help)
Level 1 1. 2. 3. etc.			
Level 2 1.			
Level 3 1.			
Masters 1.			

Getting students more involved

Although there are occasions when didactic styles may be justified — scene-setting can be important — much fieldwork now aims to get students more involved so that they learn more actively. The days of the 'Cook's Tour' in the coach or the 'lecture in the field' (HMI, 1992) are passing. We need to be careful, however, about the way in which we use local people in our fieldwork. If local key actors are going to be contacted by the students the key actors often need to be prepared. There is a danger of inviting a local official to present their case which is context-less for the student and the speaker. Should students be asked to make contact with local officials? Is there not a danger that they rush in under prepared and do themselves and our institutions a disservice by wasting the time of busy people?

There are a number of ways of ensuring that students become more involved in their fieldwork, including possibilities which might be combined in any one field exercise. These include experiential learning, role plays including public enquiries, trails and linking student fieldwork to research projects.

Experiential learning depends upon capturing the voices and experiences of local, ordinary people. It includes oral histories and interviews with local residents and people going about their everyday business (see Case Study 1).

Role plays allow students to place themselves in the role of someone who is making, or affected by, a geographical or environmental issue. The roles they play could include a developer or landowner, a national or local government officer, a pressure group or a local resident.

Mock public inquiries can provide a particularly effective role-play format within which to debate land development or other environmental issues. The public inquiry format translates readily into a fieldwork setting and operates equally well in residential or institution-based fieldwork.

The Public Inquiry format: Students are allocated roles, often in groups, of the main actors in the debate on an environmental issue. They may then collect information and data to help them take on the role, sometimes by contacting the people who carry out that role in real life. The public inquiry format requires students to make formal oral presentations and abide by a series of rules in debating the issue. The success of the debate depends on the preparation and presentations by the students but also on careful steering and control of the proceedings by someone playing the role of a Government-appointed inspector. Examples of successful use of the public enquiry format include:

1) The culmination of a six-week study on an open-cast mine proposal, which was part of a compulsory module in the final year of a course. Students worked in teams of 4 and the presentation included a 10-page typed report. The inspector was a local government officer.

2) A one-day project during a residential fieldcourse in first and second-year fieldcourses, looking at, for example, out-of-town retail development. The inquiry is held in the evening. The format is extremely flexible, but gives opportunities for students to explore roles and gain an understanding of the stance of different key actors in environmental or land development issues.

(Livingstone, 1998)

Trails provide an opportunity for a guided tour (most usually on foot) following some directed route. They can be written by the tutor for the students to follow without the tutor present, or the tutor may act as an interpreter of some of the material covered on the trail. Sometimes tutors ask students to create a trail for a particular client group which requires that students familiarise themselves with an area, collect information about the topic, and present it in some specified format. Table 1.2 (overleaf) lists some recently published examples of geographical trail exercises for students.

Linking fieldwork to research

Increasingly, taught courses are including fieldwork which can feed into wider research projects. Using students in staff research projects or consultancy work raises important ethical issues about exploitation and the level of acknowledgement of the student input required. Here are two examples (one from human and the other from physical geography) of undergraduate fieldwork which generates information and output of value beyond simply completing the project as part of a course.

Table 1.2: Some recent examples of trails

Type of trail	Style of trail	Location	Reference
Rural	tutor-led, pre-written	around Lancaster	Clark (1997)
Urban (perception studies)	student-authored trails for particular client groups e.g. partially sighted	Exeter	Dove (1997)
Urban	three, self-guided introductory trails for first-year students	Birmingham	Slater (1993) and at the GDN website (see Section 4.3)

At the University of Arizona, student fieldwork in economic geography has been linked to an ongoing research project on the Arizona Community Data Set. Small communities in the state have asked the project team to provide an analysis of the economic base of their community. This work has given students the opportunity to go into the field to collect and analyse regional economic data, and apply theories and concepts learned in the classroom. Students worked as a team to collect information usually during the summer vacation, and the results have to be presented to the community at the end of the project. Work on studies of 47 communities carried out over a twenty year period was reported by Vias (1996).

A similar approach has been taken in the preparation of information about RIGS (Regionally Important Geomorphological Sites). At the Cheltenham & Gloucester College of Higher Education students taking a geomorphology module were asked to produce reports on potential RIGS in the local area with a view to providing information to inform the designation procedure (McEwen, 1996).

Enhancing fieldwork: virtual fieldtrips

For most of us there is no substitute for the real thing. However, there are occasions when financial cost or other resourcing problems mean we cannot carry out all the fieldwork we would like to. A useful addition to the fieldwork programme, possibly in preparation for the real thing, might be for students to undertake virtual fieldwork. Virtual fieldwork uses Internet technology to make available information about field sites in the form of text and still images and sometimes as video excerpts.

The URL of the Virtual Geography Department Project at University of Texas, USA, a useful starting point for virtual fieldwork, is included in Section 4.3. There is also an address for a virtual fieldwork project, whose team consists of staff from Leicester University, Birkbeck College London and Oxford Brookes University. Both projects attempt to provide some of the experience of fieldwork via computer. Virtual fieldwork is also growing in importance in the earth sciences, and an introductory URL is provided in Section 4.3.

2.4   Cost problems: staff time and student finances

Fieldwork can be very expensive both in the use of staff time and in the financial cost. Often an increase in financial costs at a time of straitened institutional and departmental budgets has meant that the financial cost has been passed on to the student. In some institutions fieldwork has been abandoned because of these problems of cost. If we wish to maintain a fieldwork programme, we need to develop strategies to handle the pressures on time and finances.

Staff time

Fieldwork can be very costly in staff time. Either we justify that cost by arguing the returns are sufficiently great to warrant this use of staff or we develop strategies which make fieldwork less staff-intensive. There is a limit to how much staff contact can be reduced because of health and safety considerations so we have to be ready to justify heavy staff costs.

Here is one description of trends in the response to the pressures:

> *"Traditionally, field trips in higher education have consisted of small student groups receiving intensive supervision and tutoring. With very small groups it is possible for a skilful tutor to foster in the students an enquiring and reflective attitude. With larger groups, however, fieldtrips become walking lectures (or even bus tours with commentary) and students become passive."*
>
> *(Habeshaw et al., 1992, p.117)*

And here is one list of ways to cut staff costs while maintaining some of the quality (adapted from Jenkins, 1994):

- carefully cost present practice;
- stop doing fieldwork;
- give up other things to protect fieldwork;
- focus on key objectives;
- ration it carefully;
- make the ground rules and safety regulations very clear;
- make the best use of it and act to protect it;
- do (at least some of) it locally;
- watch the assessment;
- use other resources, such as self-guided trails, to replace (some of) the fieldwork;
- don't just rely on academic staff;
- research and document what is effective and efficient.

Student finances

As institutions are less able to carry the cost of fieldwork, we often increase the burden on students. In the UK this increasing burden comes at a time when student finances are being further stretched by diminishing maintenance grants, a move to a student loan scheme and the introduction of student contributions to tuition fees. Additionally, a greater proportion of students now enter HE as 'mature' students and therefore have domestic commitments which make extended periods of residential fieldwork difficult. We need to ensure that fieldwork is worthwhile because of both the financial and the time expenditure we expect of students.

Some institutions have acknowledged the problems of student commitments of time and money by offering a choice of locations. At Liverpool John Moores University students take a compulsory fieldcourse in the final year of their geography degree course. The fieldcourses are part of modules covering the USA and Canada, China and the City in Europe. Students are given a voucher covering most of the cost of the European fieldtrip: if they want to take a more expensive option they must supplement the voucher to cover the additional cost. Before following this route we should consider whether we are really providing the same learning experience to students whose fieldwork is undertaken in, for instance, southeast Asia or southern Britain. There are issues of equity and fairness raised here: should there be a range of locations offered to students providing them with different cost options?

A pool of fieldwork manuals

One way to cut costs in staff time would be to cut time spent in preparation and this could be achieved by pooling resources. There is now a huge collective body of knowledge within UK higher education, both about undertaking fieldwork and about individual field locations. A national pool of fieldwork manuals might ease the burden of preparation (a similar approach has been suggested for questions for assessment by objective testing).

The advantages of pooling documentation about fieldwork could be:

- saves time searching the literature and preparing materials;
- exchanging material leads to an incremental increase of the knowledge about a location or site;
- much fieldwork is about methodology and approach rather than the specific location so 'pooling' makes sense.

The disadvantages might include:

- a feeling of loss of autonomy by staff;
- a fear that someone might be getting something for nothing by benefiting from our effort (essentially plagiarism);
- a concern that any convergence of what is taught smacks of an higher education national curriculum by stealth;
- an increase of pressure on popular sites;
- fieldwork becomes bolted onto rather than an integrated part of the course;
- what makes sense in one course may not fit into another.

Some (fairly informal) exchange of information has already taken place on the email discussion list, GeogNet, set up specifically for discussion of issues associated with teaching geography in higher education (see Section 4.4 for more information on GeogNet).

Consider the ways in which we can make use of existing resources to avoid "reinventing the wheel" while overcoming the disadvantages.

2.5 Choosing locations (and reducing the burden on popular fieldwork locations)

With increasing student numbers (can you beat Plymouth's fieldtrip to Ireland for 200+ students?) and concerns about sustainable tourism, we need to be careful about the pressure placed on popular fieldwork locations. Much school fieldwork in the past has taken students to see the textbook examples they have read about. Pressure on popular sites implies that students have to be taken to sites which exemplify some model or ideal. As more fieldwork becomes about 'the ordinary', the pressure on sites lessens. None the less, especially with fieldcourses abroad, we need to be sensitive about taking large groups of students, sometimes predominantly 18-21 year olds, into foreign locations.

Fieldwork abroad

Paradoxically, as we move to more individual funding of fieldwork, locations further afield become possible, at least for some students. Fieldtrips to, for example, Hong Kong, Australia or Las Vegas, while still exceptional, are being introduced into some courses. Students who do participate in optional overseas fieldcourses are often very enthusiastic about the experience. For many courses, a fieldcourse abroad is a major element of their marketing strategy and may be retained for this above all other reasons.

Some of the questions which arise from fieldwork overseas include: does fieldwork abroad justify the expense? Can language problems be sufficiently overcome to enable data collection, particularly in human geography? What are the issues about transporting appropriate equipment? What about local links? Should we put some/more effort into making links with local universities?

2.6 Issues of equal opportunity

As with other aspects of the curriculum, we need to be concerned with what we are teaching and for whom. Rose (1993) suggested that we should mistrust much fieldwork because it too often displays the following traits:

- a disciplinary history which has emphasised heroic deeds at the expense of other narratives;

- its exoticisation of the unfamiliar and feminisation of the natural;

- its reliance upon, and promotion of, an archimedian, masculine perspective (or gaze), both in the 'expert' knowledge 'explaining' and defining the 'field' and in its reliance upon the objectifying techniques of field sketches and surveys;

- its parallel, though contradictory, privileging of (the pedagogue's) 'authentic' experience over intellect;

- its social practice which is often regarded as deeply misogynistic, built upon a culture of masculinism and excessive drinking.

Although not everyone will identify with or agree with all of these points, there is widespread concern, particularly among social and cultural geographers, about these issues.

> At the University of Brighton staff have begun to try to address these concerns by introducing seeing (actually being there), looking (being critical about what you see) and feeling (developing a sense of place through engaging with the location) into an overseas fieldcourse. The outcome for them is that students: develop a stronger sense of themselves, both as individuals and as members of a group; are empowered to draw on their own thinking and experience to question received academic wisdom; debate with each other; and increasingly see learning as an active, group process.

> Consider the extent to which the traits listed by Rose manifest themselves in your fieldwork programme. Is this an issue you want to confront? If so, how might you do this?

2.7   Assessing fieldwork

General approaches to assessment are covered in the Guide on "Assessment in Geography" (Bradford & O'Connell, 1998) in this series. The issue here is that the assessment of the fieldcourse needs to be appropriate to the style of teaching and learning and should be weighted appropriately for the amount of effort that is expected.

How much is it worth?

There are some critical questions about how much fieldwork counts in the overall assessment of the course. If we are going to expend considerable time, effort and money on a fieldcourse, this should be reflected in its relative weighting. Does one week's residential fieldcourse merit its assessment as a full module in the degree scheme?

How should we assess fieldwork?

> The key question here is are there features of the fieldcourse which mean that it should or can be assessed in a different way from other parts of the course?

By its nature the fieldcourse lends itself to some self-contained or stand-alone assessment strategy. Frequently it involves group work and sometimes presentations. A key question is the amount of work which can be completed within the time-span of the fieldcourse. The traditional approach has been to look for reports to be written up on the return from a fieldcourse. There are some advantages to having all (or nearly all) the work completed during the course.

The list below covers some of the advantages of each approach. Often the advantage of one is a disadvantage of the other. We should consider which of these approaches is most appropriate to the fieldwork we run.

Advantages of self-contained fieldwork:

- fresh in the mind;

- gives the projects undertaken on the fieldcourse a sense of 'completeness';

- can allow students to report on their work at the location of the research;

- does not hang over students.

Advantages of follow-up assignments:

- give students time to reflect on what they have learnt;

- supports what they have learnt with further reading;

- undertake more sophisticated data analysis;

- ensure better links with the rest of the course.

2.8   Where does it fit in the curriculum?

One key element of successful fieldwork is ensuring that it fits into the wider course curriculum. A common pattern in the past has been to run a stand-alone residential fieldcourse in each year of the degree course. Frequently, although not always, these courses had little explicit relationship to the other elements of the teaching programme such as the lecture courses or practical classes. Increasingly, geography course teams have moved to integrate the fieldwork with other elements of the degree programme and have tried to demonstrate that the fieldwork fulfils some of the objectives or delivers specific outcomes of the course. For some course teams this means that the fieldwork is one element of a module of study, itself a portion of the overall programme. Fieldwork is preceded by lectures and assignments, including plans for individual work while abroad. In some cases students are asked to put together a guide to the area which they will be studying in advance of the fieldcourse (for example, Bradbeer, 1996). Fieldwork is increasingly incorporated into modules on geographical techniques at level 1 or 2 (first or second year) as early preparation for a final year dissertation and into more theoretically-based modules on, for instance, environmental issues and values later in the course, often at level 3 (third year).

For fieldwork to continue to be of value in the curriculum of geography higher education courses we will need to justify it as an integral part of the overall programme. There are two good reasons for this. One is that it is pedagogically much sounder if the programme of study which students are following is coherent with an overall set of aims. More pragmatically, it will be more difficult for resource managers to argue that fieldwork is an unnecessary extra if we can demonstrate that it is fully integrated into the curriculum and fulfils many, or at least some, of the key objectives of our courses.

3 Case studies

3.1   Case study 1

Title: Streetwork: an encounter with place

Originator: Jacquelin Burgess and Peter Jackson

Contact: Department of Geography, University College London, 26 Bedford Way, London, WC1H 0AP, UK.

[NB Peter Jackson is now at Department of Geography, University of Sheffield, and uses this field exercise with students there.]

Reference: *Journal of Geography in Higher Education*, (1992) 16(2), 151-157.

Keywords: streetwork, qualitative geography, group work

This fieldwork forms the core of a second level course in cultural geography at UCL. The objective of the project is to get students to: encounter a place with which they are unfamiliar; 'open up' to the urban experience; and describe and interpret symbols and meanings that are conveyed through that experience. At the end of the fieldwork all students produce an interpretative essay of 2,500 words which conveys their experiences as both travellers and explorers. Students work in groups of no more that four and provide a progress report at mid-stage to the rest of the class.

The location should be somewhere that the group does not know well. The place could be a nineteenth century High Street, an Edwardian suburb, a tenement block, a modern housing estate, a gentrified neighbourhood or an ethnic enclave.

Within their selected environment students should take time to become aware of the place by watching, looking and listening, but much more acutely than normal. In the course of their fieldwork students should:

- talk to people, travel on local transport, buy things, and ask for directions;

- contact someone who is knowledgeable about the area and who can give an introduction (a community worker, a teacher);

- take photographs, make sketches, draw maps and keep a diary;

- find out about the area's history;

- find out what issues concern people today.

It is up to the group to organise their own time and to follow up their own leads.

Recipe for case study 1

Level: all levels. Location: urban locations. Examples are from UK and US. *Class size:* not important. Students work in groups of 4. *Timetable:* either intensively on a residential fieldcourse or as a more extended project linked to coursework.

Figure 1.1: *Streetwork: student handout*

The objective of this project is to *encounter a place* with which you are currently unfamiliar; to open yourself up to the urban experience; and to describe and interpret the symbols and meanings that are conveyed through that experience. As cultural geographers, you will need to question the extent to which 'people and place' are indivisible. To begin with, you will be an 'outsider', open to features that may have become commonplace or routine for local people. Your perceptions may be more acute that an 'insider's' less focused curiosity, dulled by routine observation and habitual experience. But you may be unaware of the subtle nuances of meaning that structure and hold communities together in place. Take your time to get used to the area and don't make hasty judgements.

The aim is to produce an *interpretive account* of your chosen place, conveying your experiences as a traveller and explorer. Think carefully about the language you will use, the analysis you will make.

Where?   Choose somewhere you don't know well. Spend time walking around until somewhere 'feels' interesting. Exercise you intuition and empathy (you've all got it!). The place could be a nineteenth-century high street, an Edwardian suburb, a 1930s shopping centre, an industrial backstreet, a tenement block, a market, a modern housing estate, a major development (like Barbican), a gentrifying neighbourhood, or an 'ethnic' enclave (like Brick Lane).

How?   Become *aware* of your environment: watch, look and listen more acutely than normal. Concentrate on specific features of the environment (colours, sounds, faces, architectural styles, graffiti, street furniture, clothes...): whatever makes the area different or distinctive.

- talk to people, on buses, buy things, or ask directions.

- contact someone who is knowledgeable about the area and who can give you an introduction (a community worker, a residents' association, a local teacher).

- take photographs, make sketches, draw maps and keep a diary.

- collect information about the area, read the local paper, use the library, visit the parks.

- find out about the area's history, about important local buildings or famous residents.

- for the times beyond loving memory, examine the character of the built environment, historic institutions.

- find out what issues concern people today.

You will be working in *groups of four people*, so you should try to work out a convenient division of labour. But, at least for the first few times, visit the area as a group. Compare your impressions of the neighbourhood and decide who will follow up which leads. Each group will be asked to make a *progress report* to the rest of class. By then you should have chosen an area, begun the research and thought about a possible theme. The *end product* will be an interpretive essay of 2500 words with appropriate illustrations.

3.2   Case study 2

Title: Student-authored fieldtrails

Originator: David Higgitt

Contact: Department of Geography, University of Durham, Science Laboratories,
                 South Road Durham, DH1 3LE, UK. Email: d.l.higgitt@durham.ac.uk

Reference: *Journal of Geography in Higher Education*, (1996) 20(1), 35-44.

Keywords: geomorphology, trail, peer assessment, self assessment

This fieldwork formed part of a second level course in geomorphology (although it could be extended to other parts of the geography curriculum including human geography). While the course aims to introduce and reinforce themes for understanding and interpreting landforms in general, its execution and associated coursework are concentrated in a specific locality. The fieldwork involves students working in teams in an area close to the university to produce geomorphological trails and information boards targeted at a designated audience (upper school or first year undergraduate geographers). The fieldwork sessions last 3-4 hours over a five week period with two further weeks to hand work in.

Figure 1.2: A timetable of class and assessment activities

Week	Class activity	Assessment activity
1	Background to local geomorphology Introduction to trail design Definition of groups and tasks Teamwork activity Define peer assessment criteria Establish possible routes Arrange meetings outside class	
2	Field 'lecture' Initial work on trail design	
3	Trail design	
4	Group A guides others along part of trail	Teacher assessment of input of individual Group A students Group B provides written feedback (assessed by teacher)
5	Group B guides others along part of trail	Teacher assessment of input of individual Group B students Group A provides written feedback (assessed by teacher)
6		
7		Submission of group work Submission of individual summaries Submission of peer assessment scores

Other papers concerned with using trails in fieldwork are listed in Section 4.

Recipe for case study 2

Level: second level but could be adapted for other levels. *Location:* example is from rural location in NW England. *Class size:* not important, students work in groups of 4. *Timetable:* in this example, five 2-hour sessions related to course work, but could operate on a residential fieldcourse. *Total hours:* 10 contact hours.

3.3   Case study 3

Title: Self-paced distance learning packages for large group fieldwork

Originator: Peter Keene

Contact: Geography Unit, Oxford Brookes University, Oxford, OX3 0BP, UK.

Reference: *Journal of Geography in Higher Education*, (1993) 17(2), p.159.

Keywords: distance learning, trail, manual

Problem: ***too many students, too few staff***

Teachers are facilitators, encouraging students towards autonomous learning (where sufficiently motivated students have acquired the skills to control and direct their own education). Some students achieve autonomous learning status early, others require greater tutor assistance (staff time). Traditional tertiary teaching methods have favoured tutor-orientated programmes where a favourable SSR (student:staff ratio) is critical. The success of this system is threatened by increasingly large classes, an immediate effect of which is to decrease the individual contact time between staff and student in the field.

Response: ***distance learning packages for self-paced fieldwork***

If your field classes consist of open-air lectures, then larger groups probably present few difficulties other than hoarseness! However, if you embrace the tradition of problem-solving activities with small groups guided by the tutor, then larger groups pose new problems of tutor availability. One strategy for the more effective use of declining staff availability is to provide better structured distance learning packages (material which encourages work undertaken without the direct supervision of the tutor).

Shrewdly written, these packages can provide reassurance and support for hesitant students without recourse to the tutor. Tutor time can thus be conserved either for those who are outstripping the scope of the package, or for those who need additional assistance to accomplish the task. Such packages might include reference material, written information, instructions, guidance on specific exercises, manuals, guides or trails. Three approaches are summarised below.

Main Features

Manuals: Non site-specific reference packages providing information on a 'need to know' basis, to enable students to interpret a specific site. An 'interactive' manual encourages student participation in problem solving.

Trails: A booklet, usually assuming some thematic progression and no tutor presence, linking a series of sites selected for study. Attributes to be considered in making trails suitable as a student exercise include:

- Instructions clearly explaining route and the nature of the exercise (not necessarily prescriptive directions).

- Stimulation through questions, tasks and comment to encourage active participation and self-directed investigation.

- Reference material built into the trail or in manual form and availability of appropriate instrumentation. All to provide an exercise sufficiently clearly explained and structured to give support and reassurance to weaker students whilst not constraining originality and deeper investigations by stronger students.

Student produces material: A successful alternative is to turn the tables completely so that it is the students themselves who are preparing the guides, trails or manuals for a specified target audience.

Gains and losses

Plus Points:

- Although they do not save tutor time, they do redistribute it, giving more time in the field to deal with problems.

- It is tutor-stimulating, focusing attention on central needs of students as they advance towards self-paced, autonomous learning.

- Student preparation of packages (trails, guides, manuals) is a valid and stimulating exercise in its own right and is a suitable fieldwork task for students, emphasising problems of selection, explanation and communication.

- Early introduction to self-paced, distance learning packages underscores the route to autonomous learning. They reflect positive advantages in throwing learning responsibility onto students whilst at the same time providing a structured safety net.

- A published trail or manual is capable of reaching a large extra-mural audience.

Minus Points:

- Distance learning packages are only appropriate for part of the student's educational experience.

- They are not a good substitute for stimulating interaction with a perceptive tutor.

- Without care they can encourage a low standard deviation in performance, supporting the weak, but confining the high-flier.

- Consideration needs to be given to the safety of students when not under immediate supervision.

- It is not a way of saving staff time. Preparing packages is time consuming.

Relevant References

Manuals   Keene, P. (1982)

Trails   Keene, P. (1989)

Students   Peterson, J.F. (1984)

4 References and other sources

4.1 References cited in the text

Bradbeer, J. (1996) Problem-based learning and fieldwork: a better method of preparation? *Journal of Geography in Higher Education*, 20(1), pp.11-18.

Burgess, J. & Jackson, P. (1992) Streetwork — an encounter with place, *Journal of Geography in Higher Education*, 16(2), pp.151-157.

(See Case Study 3.1)

Gold, J.R., Jenkins, A., Lee, R., Monk, J., Riley, J., Shepherd, I. & Unwin, D. (1991) *Teaching Geography in Higher Education* (Oxford: Blackwell).

Chapter 3 is on fieldwork. They bemoaned the lack of material on fieldwork. This gap in the literature largely remains although more case studies have been published in the *Journal of Geography in Higher Education*.

Habeshaw, S., Gibbs, G., & Habeshaw, T. (1992) *53 problems with large classes: making the best of a bad job* (Exeter: BPCC Wheatons).

Higgitt, D. (1996) The effectiveness of student-authored field trails as a means of enhancing geomorphological interpretation, *Journal of Geography in Higher Education*, 20(1), pp.35-44.

(See Case Study 3.2)

Jenkins, A. (1994) Thirteen ways of doing fieldwork with large classes/more students, *Journal of Geography in Higher Education*, 18(2), pp.143-154.

Keene, P. (1982) The examination of exposures of Pleistocene sediments in the field: a self-paced exercise, *Journal of Geography in Higher Education*, 6(2), p.109-121.

Keene, P. (1989) Trails on trial, *Environmental Interpretation*, 44, pp.15-16.

Keene, P. (1993) Self-paced distance learning packages for large group fieldwork, *Journal of Geography in Higher Education*, 17(2), p.159.

Kent, M., Gilbertson, D.D. & Hunt, C.O. (1997) Fieldwork in geography teaching: a critical review of the literature and approaches, *Journal of Geography in Higher Education*, 21(3), pp.313-332.

Livingstone, I. (1998) Role-play planning public inquiries, *Journal of Geography in Higher Education*, submitted.

McEwen, L. (1996) Student involvement with the Regionally Important Geomorphological Site (RIGS) Scheme: an opportunity to learn geomorphology and gain transferable skills, *Journal of Geography in Higher Education*, 20(3), pp.367-378.

Peterson, J.F. (1984) Preparing environmental interpretation literature, a strategy for undergraduate teaching, *Journal of Geography*, April, pp.73-78

Rose, G. (1993) *Feminism and geography: the limits of geographical knowledge* (Cambridge: Polity Press).

Vias, A. (1996) The Arizona community data set: a long-term project for education and research in economic geography, *Journal of Geography in Higher Education*, 20(2), pp.243-258.

Warburton, J. & Higgitt, M. (1997) Improving the preparation for fieldwork with 'IT': two examples from physical geography, *Journal of Geography in Higher Education*, 21(3), pp.333-347.

> The two examples are using CAL to introduce concepts and using a GIS to provide background information about the field locality. (This article is available in an on-line version of the *Journal of Geography in Higher Education*. Information about accessing the article can be found at http://www.carfax.co.uk/cfx-elec.htm.)

4.2   Other sources

Some abstracts covering fieldwork (and other topics) can be found at the Geography Discipline Network WWW pages:

> http://www.chelt.ac.uk/gdn/

The following is a selective list of recent papers covering fieldwork:

Clark, G. (1997) The educational value of the rural trail: a short walk in the Lancashire countryside, *Journal of Geography in Higher Education*, 21(3), pp.349-362.

Dove, J. (1997) Perceptual geography through urban trails, *Journal of Geography in Higher Education*, 21(1), pp.79-88.

Ellis, B. (1993) Introducing humanistic geography through fieldwork, *Journal of Geography in Higher Education*, 17(2), pp.131-139.

Gemmell, A.M.D. (1995) 'Competitive' simulation in the teaching of applied geomorphology: an experiment, *Journal of Geography in Higher Education*, 19(1), 29-39.

Gold, J.R. & Haigh, M.J. (1992) Over the hills and far away: retaining field study experience despite larger classes, in: G. Gibbs & A. Jenkins (Eds) *Teaching Large Classes in Higher Education: how to maintain quality with reduced resources*, pp.117-129 (London: Kogan Page).

Haigh, M. & Gold, J.R. (1993) The problems with fieldwork: a group-based approach towards integrating fieldwork into the undergraduate geography curriculum, *Journal of Geography in Higher Education*, 17(1), pp.21-32.

Haigh, M.J., Revill, G. & Gold, J.R. (1995) The landscape assay: exploring pluralism in environmental interpretation, *Journal of Geography in Higher Education*, 19(1), pp.41-55.

Her Majesty's Inspectorate (1992) *A survey of geography fieldwork in British degree courses, Summer 1990-Summer 1991, Report 9/92/NS* (Stanmore, HMI Department of Education and Science). extracted in *Journal of Geography in Higher Education* (1993), 17(1), pp.35-36.

Jenkins, A. (1997) *Teaching more students: fieldwork with more students* (Oxford: Oxford Centre for Staff Development)

> Lots of practical suggestions for dealing with large classes on fieldwork.

Katz, C. (1994) Playing the field — questions of fieldwork in geography, *Professional Geographer*, 46, pp.67-72.

Kneale, P. (1996) Organising student-centred group fieldwork and presentations, *Journal of Geography in Higher Education*, 20(1), pp.65-74.

Kobayashi, A. (1994) Coloring the field: gender, "race" and the politics of fieldwork, *Professional Geographer*, 46, pp.73-80.

Lewis, S. & Mills, C. 1995. Field notebooks: a student's guide, *Journal of Geography in Higher Education*, 19(1), pp.111-114.

Lonergan, N. & Andresen, L.W. (1988) Field-based education: some theoretical considerations, *Higher Education Research and Development*, 7, pp.63-77.

Madge, C. (1994) 'Gendering space': a first year geography fieldwork exercise, *Geography*, 79, pp.330-338.

McEwen, L. (1996) Fieldwork in the undergraduate geography programme: challenges and changes, *Journal of Geography in Higher Education*, 20(3) pp.379-384.

> This is the introductory paper for a set edited by Lindsey McEwen from a symposium. It addresses some of the issues also raised here.

Mosser, J. (1995) Participatory student field guides and excursions, *Journal of Geography in Higher Education*, 19(1), pp.83-90.

Nast, H. (1994) Women in the field: critical feminist methodologies and theoretical perspectives, *Professional Geographer*, 46, pp.54-66.

Orion, N. & Hofstein, A. (1994) Factors that influence learning during a scientific field trip in a natural environment, *Journal of Research in Science Teaching*, 31, 10, 1097-1119.

Slater, T.R. (1993) Locality-based studies and the Enterprise Initiative, *Journal of Geography in Higher Education*, 17(1), pp.47-55.

> Describes an introductory module for first-year students which includes lectures and self-guided trails covering human geography and planning issues in the area around the university campus (Birmingham). (See GDN Web pages, http://www.chelt.ac.uk/gdn, for a fuller description).

Tinsley, H.M. (1996) Training undergraduates for self-directed field research projects in physical geography: problems and possible solutions, *Journal of Geography in Higher Education*, 20(1), pp.55-64.

4.3   CAL (Computer Aided Learning) and virtual fieldwork

In the UK the Computer Teaching Initiative (CTI) Centre for Geography, Geology and Meteorology (CTIGGM) based in Leicester provides a starting point for finding geographical web resources:

> http://www.geog.le.ac.uk/cti/

There is also a CTI Centre for Land-Use and Environmental Sciences (CLUES):

> http://www.clues.abdn.ac.uk:8080/

There are a number of CAL packages which are appropriate for fieldwork, some of which were produced by HEFCE's Teaching and Learning Technology programme. In the TLTP's GeographyCAL package there is a "Social Survey Design Tutorial", a description of which can be found at the CTIGGM WWW site:

> http://www.geog.le.ac.uk/cti/Tltp/tech.htm

The TLTP earth sciences consortium has descriptions of modules on "Preparing for fieldwork: using a compass clinometer" and "Geological map skills" on their WWW site:

http://www.man.ac.uk/Geology/CAL/modules.html

There is an interactive surveying tutorial at the Project Interact site:

http://www-interact.eng.cam.ac.uk/survey/index.html

There are a number of worldwide web sites offering virtual fieldwork, most of which are based in the USA. The Virtual Geography Department Project at the University of Texas, which includes virtual fieldwork, is at:

http://www.utexas.edu/depts/grg/virtdept/contents.html

The virtual fieldwork project run by Leicester University, Birkbeck College London and Oxford Brookes University, can be found at:

http://www.geog.le.ac.uk/vfc/

A directory of sites for virtual fieldwork in the earth sciences can be found at:

http://www.uh.edu/~jbutler/anon/anontrips.html

4.4   Email discussion: GeogNet

GeogNet is an email discussion list set up specifically to deal with issues arising from teaching geography in higher education. It is used predominantly by teaching staff wanting help with curriculum design and course delivery, but also carries other information such as conference notices and occasional job adverts. Contact GeogNet by emailing GeogNet@nene.ac.uk.

Unit 2: Dissertations

1 Introduction

This unit of the Guide looks at dissertations within the geography curriculum. In Sections 1 and 2 we start by asking whether the dissertation fulfils the aims of our courses in the most effective and efficient way. The traditional role of dissertations in the geography curriculum is increasingly being reconsidered. One reason for this is that growing numbers of students in higher education make it more difficult for staff to support and supervise lengthy individual projects. Additionally recent debate has focused on the extent to which higher education courses prepare graduates for work or for lifelong learning. Does the dissertation best fulfil these aims? Section 3 explores issues of managing the dissertation process and Section 4 suggests some sources of further information. We will be looking at several examples of good practice, and at ideas about dissertations. You may not agree with some examples of existing practice, or some of the ideas expressed here. You should see them as a mirror in which to review your own courses and teaching.

Here are some responses from a discussion on dissertations from the GeogNet email discussion list (Email: GeogNet@nene.ac.uk). They were prompted by a contribution from a course team who were thinking of making dissertations optional and wanted advice, opinions or experience.

"We made our undergraduate dissertations optional this year for the first time. ...There was considerable resistance from some colleagues but basically in view of huge marking and supervision loads the option lobby won. ...Now all (level 2) students have to undertake a Research Proposal module (10 credits) and they then choose if they want to go on to do a dissertation (40 level 3 credits). ...students have to do an extended essay if they choose not to do a dissertation."

"Since March 1995 the dissertation... has been optional."

"We have INCREASED the weighting on the dissertation. We regard it as a more meaningful, and career-relevant, test of a student's ability... I think any degree is severely undervalued if no dissertation is included."

"Our dissertation is optional and students not doing it can still get an honours degree. Personally, I do not like this and feel that an important element of the training is lost."

"...being asked to present a coherent and well structured report is something they may well find themselves asked to do (after they leave university). Their dissertation not only demonstrates that they can DO geography as well as write about it but gives them practice in a skill that they may well need to use throughout their lives. I would argue that we would do our students a great disservice both in terms of their training as geographers and development of transferable skills if we did not require a dissertation as part of their undergraduate course."

2 Dissertations: pros and cons

2.1 Why do dissertations?

The appropriateness of the dissertation in the geography undergraduate curriculum is the subject of some debate. Here are some of the reasons why dissertations have been (and still are) a significant component of final year work in many geography courses.

- *Independence and originality*

  One aim of the undergraduate curriculum should be to help students to become independent learners. Unlike much of the rest of the course students can select and carry out the study according to their own agenda, albeit within a negotiated framework. The dissertation is probably the best opportunity for undergraduate students to demonstrate some independence and originality in their undergraduate studies. Students have the opportunity to see a project through from its initial inception to its completion as the submitted dissertation.

- *Organisational skills*

  Independent inquiry involves and develops the ability to plan, consult, negotiate, implement, schedule and deliver.

  "It helped me to see I should develop organisational skills, but it didn't help me to do so, at all, since it was left up to us, mostly." — Ruth (graduate student)

- *Application of knowledge and skills*

  The dissertation allows students to demonstrate their ability to apply the knowledge and skills they have learned in other parts of their course.

- *Deep learning*

  When students have the freedom and responsibility to choose a topic which they will research, provided the process of choosing the topic has been successful, they are likely to be highly committed to it. Added to this, the dissertation student should be engaging in the highest levels of thinking. The likelihood that they will engage with the topic in terms of deep and long term learning is therefore particularly strong.

  "The choice is crucial, not the length. If you're into it, you'll learn from it. If not, you'll be counting the words all the time." — Chris (graduate student)

2.2   Problems with dissertations

We might see the list in Section 2.1 — independence and originality, organisational skills, application of knowledge and skills, and deep learning — as intended and desirable outcomes of the undergraduate dissertation. However, there are a number of pitfalls which can militate against the achievement of these.

- *Is it independent?*

  Students may be over-reliant on the advice of their supervisor or follow paths trodden by many other students when choosing a topic to investigate. In addition, nowadays there is much technology available, including the World Wide Web and optical scanners, which enables the less committed students to plagiarise and incorporate chunks of undigested material into their work.

- *Are they organised?*

  Independent inquiry is difficult to formulate, negotiate, plan, carry out and deliver on schedule. Without guidance, support and exemplification of good organisational practice, the dissertation experience may have only limited impact in helping students develop these skills. If curriculum time is not explicitly devoted to developing these skills, students may not automatically exhibit them when conducting their investigation.

- *Do they apply knowledge and skills?*

  Giving students too much independence may lead to a dissertation topic, chosen by the student, which is not directly linked to areas already studied in the undergraduate programme. In this case it is less likely that there will be a clear progression from the acquisition of skills and knowledge in the earlier part of the course to their application in the dissertation.

- *Do they demonstrate deep learning?*

  Given the level of independence and autonomy that students are given when undertaking a dissertation, there is a danger that the academically weak student will fill the allotted wordage with too much description, statements of basic principles, or unanalytical literature review. To have value in the geography curriculum, dissertations need to move beyond descriptive narratives.

Consider the following statement:

"If we are preparing students for the world of work and for life-long learning the traditional dissertation is a poor preparation. It relies on long-term, sustained effort (much of it away from the desk), individual effort and the production of a lengthy piece of prose. Students could be using their time better."

What are the reasons for doing dissertations. Are there better ways of achieving course aims than by dissertation?

2.3   Where do dissertations fit in the curriculum?

"Our research skills and report writing module was useful when it came to the dissertation — and still is." — Ruth

> The traditional place of the dissertation in the curriculum is under review in some geography courses. The key questions are: Is the dissertation essential to a geography degree course? Could it be optional or dropped altogether? Why?

The dissertation holds a special place in many courses but we need to think about how it fits into the curriculum as a whole. A common view is that the dissertation should build on knowledge and skills from elsewhere in the course, stand alone as a task, and come in the final year (or level 3) of the course. However, the comments in Section 1 of this Unit show that this view is being challenged in some courses.

The Enterprise Dissertation

One example of the change in thinking about dissertations is the Enterprise Dissertation experiment at Lancaster University. In this kind of exercise value is given to enquiry which has current practical application as well as being valid intellectual inquiry.

> Initiated through the government-funded 'Enterprise in Higher Education' scheme (EHE), the *Enterprise Dissertation* is an optional variation on the more traditional style of final year project, in that it requires students to establish a research contract with an external 'client' organisation. Successful links have been made with planners, voluntary bodies and conservation groups who often appreciate the value of research and are able to identify topics of use to them which are feasible for a student.
>
> Further information is available on the GDN Web pages at:
>
> http://www.chelt.ac.uk/gdn/abstracts/a1.htm
>
> and from:
>
> Clark, G. (1991) Enterprise education in geography at Lancaster, *Journal of Geography in Higher Education*, 15(1), pp.49-56
>
> Clark, G. (1995) Enterprise dissertations revisited, *Journal of Geography in Higher Education,* 19(2), pp.207-211

There is no perfect model of when dissertations should be completed: different courses use different patterns. These range from those which see the dissertation as something which is contained entirely within a third year programme, commencing in September and completing after Easter, to those which start their dissertation programme in year 2 with a hand-in date before Christmas.

Is there a good reason why the dissertation needs to be one of the last pieces of work that a student completes on their course? How long does it take to complete a successful dissertation? What skills and information do students need before they can start work on the dissertation? Does this affect its scheduling? Could something come after it and build on it?

Does the existing curriculum prepare students to do a good job on their dissertation?

Most geography courses place the dissertation in the second half of the degree programme in which case there is an expectation that there are skills and information which students need from earlier parts of the course before they can start.

Consider the following questions

When in the programme will students:

- have covered enough geography to yield fruitful dissertation topics?

- have developed the necessary organisational skills?

- have developed the necessary inquiry skills?

- have developed the necessary research skills?

- have developed the necessary writing skills?

- be sufficiently interested and motivated?

To what extent would you wish to structure the earlier part of the curriculum, such as research and report-writing modules, for instance, with the dissertation in mind?

3 Managing the dissertation process

Having decided to retain dissertation work in some form, at some point in the course, we need to consider how best to manage the process. This section explores the relevant issues.

3.1 Supporting students

There is a range of ways in which students can be supported while they are working on their dissertation. These include tutorial supervision, seminar programmes, documentation in the form of a dissertation guide and learning agreements. Each of these is considered briefly here. You might wish to think of additions to this list, and to discuss the relative merits of each.

Beliefs and practice differ on the extent to which students should be prepared for and supported in their dissertation work. Here are two contrasting views.

a) As maturing people, the responsibility of developing an interest, and productively researching that interest is the responsibility entirely of the student;

b) Student learning is enhanced through supervision, which need not detract from the independent nature of the enquiry.

Which of the above views most closely approximates to your departmental practice, and why?

The dissertation guide

Some geography courses produce a written guide for students (and some a guide for staff: see Section 3.3). Where it exists for students it is a key document and usually contains information to help them to do their dissertation successfully. Guides vary greatly in style. Some concentrate on the course's rules and regulations associated with the dissertation while others are more concerned with giving some pointers for producing a good piece of work. Table 2.1 (overleaf) provides a compendium of section headings taken from a number of guides for students.

Why do students need a guide? Can the information be delivered in different ways? How does your dissertation guide (if you have one) compare with the compendium list in Table 2.1?

Table 2.1: Compendium of headings taken from a number of Dissertation Guides for students

Introduction:
> Definition of a dissertation
> Why do a dissertation?

Timetable for progress and presentation:
> Introductory seminar
> Consultation with tutors for supervision
> Research of possible topics
> Initial title approval and registration
> Research to refine/redefine the topic
> Submission of the proposal
> Reallocation of supervisors (if necessary)
> Investigation
> Scheduled meetings with supervisors
> Interim reports/drafts
> Completion of final draft
> Submission of work

Choosing a topic:
> Interest
> Problem/hypothesis
> Feasibility (scope, access, ethical issues)

Making the proposal:
> Title
> Contextualisation
> Aims and method
> Action plan
> References
> Supervisor
> Equipment/laboratory required

Mutual expectations/obligations of supervisor/supervisee:
> Learning contract
> Progress proforma

Writing up:
> Layout
> Technical matters (data presentation, style, cartography etc.)
> Appendices
> Referencing

> Plagiarism

Assessment:
> Process and criteria

Seminar programme

Another way to prepare and support students is to hold seminars or tutorials. In some cases a dissertation guide will be used as the focus for an introductory seminar to initiate students into the dissertation process, but there is a range of other reasons for holding seminars.

Early seminars might:

- focus on organisational skills;

- highlight the timetable of stages in dissertation preparation and the date by which each stage should be complete;

- present and explore different sources of information to help select a topic;

- discuss issues associated with information or data collection and fieldwork;

- introduce students to staff's research interests;

- give an indication of what is expected by reading and/or commenting on past (anonymised) dissertations;

- reiterate appropriate points from the guide.

Later seminars might:

- focus on presentational skills;

- allow students to report on progress;

- provide an opportunity for interim assessment;

- discuss issues associated with data analysis, writing up and dissertation presentation;

- provide mutual support among students.

> Why have a seminar/tutorial programme associated with the dissertation? Should students meet in groups as well as with individual dissertation supervisors? What are the advantages of group seminars for an individual piece of work? When should the seminar programme start? What should it cover? How many seminars are needed?

For those courses whose dissertation spans much of level 3, one pattern is to hold an introductory seminar between March and July of the penultimate year of the course for dissertations due between January and April in the final year. Another seminar could be held in November of the final year at which students report on progress.

Assessing past dissertations

A useful exercise which enables students to begin to understand what is required is to get students to assess (anonymized) past dissertations, once having worked through the criteria, and perhaps discussed examples of good and bad practice. Two or more students can assess the same dissertation and then compare their assessments. The resulting discussion will help them form very useful insights into how their own material might be treated (Brown *et al.*, 1996, p.56).

Learning Agreements

"You should meet with the student and agree to write down a schedule. I didn't have much self discipline at the time, and I wasn't alone in that." — Ruth

Learning agreements or contracts are gradually being introduced into HE and other education sectors. Often they are a response to increasing numbers. They also provide some basis for discussions which may arise from student concerns over the level of support they receive, and might be used at appeals. The agreements are a formal, written statement of the mutual expectations and obligations of the student and the supervisor. This explicit statement helps to ensure that students are realistic about the level of support that they can expect and also helps to bring some equity to the level of supervision provided for each student. Because of the level of independence and autonomy involved, dissertations provide an ideal opportunity for learning contracts to be implemented, although, like other aspects of the curriculum, it may be helpful to have introduced elements of learning agreements earlier in the course. Figure 2.1 (from Coventry University) shows an example of a dissertation learning contract.

Learning contracts may be especially valuable in getting students to buy-in to the agreement, or if time precludes this, pre-printed forms may be used. Nevertheless an element of negotiation may be built in to the process, for example, the frequency and dates of meetings, and the stage of the work to have been completed by these dates.

> If you don't use them already, what would be the advantage of a dissertation learning contract for staff and for students on your course?

Figure 2.1: *Dissertation learning contract*

320/350 GEO - Geography/Recreation and the Countryside Project

Supervisory Arrangements and Learning Contract

The objective of this document is to outline what students can expect from their project supervisor and , in turn, what supervisors can expect from their students. In effect, it is a **Learning Contract** between student and supervisor; both are required to sign it as an agreed programme of work and each should retain a copy.

Supervisor Responsibilities

Students will be allocated officially to their supervisor before the end of the second week of the Autumn term. They can expect the following from their supervisor:

- advice on the identification of a specific topic and the aims and objectives of the project;
- approval of a programme of work for the double module;
- guidance on the most appropriate literature and how to conduct a literature review which places the research into its wider context;

- advice on the selection of the most appropriate methodology and techniques;

- advice on the analysis and interpretation of primary and/or secondary data;

- advice on how to plan the structure of the written project, including the layout of individual chapters;

- detailed comments on the draft copy of two chapters of the project (including grammar and style of writing), but only if submitted before the end of the Spring Term. This usually means commenting on two of the following: introduction and aims, literature review/context, methodology (but not analysis of results etc);

- to be monitored formally on the progress of their project in both the first and second terms. This could be achieved in a project seminar, attended by all students supervised by a particular member of staff, where each student gives a five-minute presentation on 'progress to date'. Unsatisfactory progress is reported to the module tutor. (NB a pass in the project is required for the award of an honours degree).

Student Responsibilities

Staff can expect the following from their project students:

- regular meetings (ie at least once every two weeks) with the student. The onus is very much upon the student to see their supervisor at regular intervals throughout the year and to keep to appointments made;

- identification of an area of 'interest' for the project;

- formulation of clear aims and submission of a first draft of one chapter by the end of the first term;

- discussion of adopted methodologies and techniques;

- progress reports on literature searches and contextual literature review;

- discussions on the contents of individual chapters and the structure/layout of the project;

- presentation of two formal 'progress reports' (one per term) on the project. Unsatisfactory progress will be reported to the module tutor (the implication for the student being the relegation from an honours to a general degree);

- submission of a draft copy of two chapters of the project by the end of the Spring Term (any project work submitted after that date will not be read by the supervisor);

- Detailed arrangements for supervision will necessarily vary according to the nature of the project. Nevertheless, there is and expectation that students will work closely with supervisors and consult them frequently. Students' attention is drawn to the 'projects noticeboard', where messages are often left by supervisors. It is possible for students to complete a dissertation rather than a project. The former is a 10,000 words review of the literature, with recommendations for future research, and does not involve the collection and analysis of primary data. A competent dissertation is not the same thing as a 'long' essay and students are advised to consult carefully with their supervisor about how to tackle a dissertation topic. Students are encouraged to undertake a project.

Finally, students are reminded that the project is a double module, with important implications for the final class of degree awarded. In view of this, an introductory lecture will be given by the module tutor on the aims and completion of 320/350 GEO; this will take place early in the first term.

Both student and supervisor should sign this document as an agreed programme of work:

Project topic: ....

Signatures: Student: ....   Supervisor: ....

3.2   Helping students to choose a dissertation topic

"I chose a subject my tutor was involved in. That was encouraging; she was learning at the same time as I was. That was motivating on both sides." — Chris

Many dissertation guides advocate three broad ways of choosing a topic:

- review specific topics, lectures, seminars and reading covered in the first part of their course, and develop a dissertation topic from an aspect of the course found to be of interest;

- review a topic or issue not necessarily covered by the course but which is none the less of interest;

- choose from a list of options designed to build on the curriculum and correspond to staff's interests.

Here is a checklist which you might encourage your students to consider:

- lecture courses — what has interested you in the course?

- staff or postgraduate research — where does the expertise of potential supervisors lie?

- previous dissertations — what have other students found interesting?

- national or local controversy — what is seen as a problem or as worth investigating?

- current journals — what are "the experts" concerned with topically?

- a visit or field trip — what personal experience can the student draw on?

- career opportunities — what career do they hope to pursue?

- friends and family — sometimes, but not necessarily, a last resort: do friends and family have useful contacts or experience which can be used as a starting point?

(adapted from Burkill & Burley, 1996, pp.432-433)

"She was firm about me knowing where I wanted to go at the proposal stage. That was very supportive in the end." — Chris

Feasibility

There are a number of factors which might make an initial dissertation proposal unfeasible (Gatrell, 1991, p.20):

- the topic may not enable the student to meet the assessment criteria;

- the topic may be too broad in conception and might consequently lack depth;

- the topic may be too large in scope, and not treatable within the required word-count;

- important data may not be available or accessible;

- there may be an ethical dimension preventing the selection of the original title.

"There needs to be some room to be flexible while you are doing it." — Ruth

3.3   Supporting staff

Just as it is important to consider the level of support provided for students, it is useful to think about how staff can be supported in supervising dissertations. Simply because some of us were thrown in at the deep end with supervision is no good reason not to provide support now (the "Junior-hospital-doctor syndrome").

Two main reasons given for providing guidance to dissertation supervisors are to:

- induct staff new to dissertation supervision into the process;

- ensure all supervising staff are working to comparable standards of supervision.

A way to provide this support is to produce a staff guide on dissertations. The contents of one guide are listed in Table 2.2 (overleaf). This is just one example. Which parts are/would be useful in your department, and which less appropriate? Why?

Table 2.2: *Contents list from a guide for dissertation tutors*

- Prior to the first contact with staff
  (induction of students into the process)

- First meeting
  (Timing, short, informal)

- Number of students per staff member
  (five from each of two courses)

- Decision on which students to supervise
  (When the decision is made; with what information)

- Non-attached students
  (will be allocated by the second week of the final year at the latest)

- Number of contacts with students
  (minimum four; no maximum; guard against overdependence; be wary of students
  appearing for as few as two meetings; plagiarism)

- Duration of meetings
  (to complete the aims identified at the beginning)

- Nature of discussion
  (generally student-led; supervisor's contribution gradually reducing; advise and
  suggest; agree clear goals at the end; students' demands should not be excessive)

- Before students collect data
  (approval of proposal first)

- Policy on drafts
  (refuse drafts submitted too close to a deadline; read any section only once; offer
  general advice only. Do not: outline the exact method for conducting an
  investigation; calculate statistical analyses; lead the interpretation of the results)

- Log books
  (show tasks completed and planned, agreed goals, issues not settled design
  suggestions etc. Read, sign and date. Specify next meeting)

- Policy on extensions

- Poor attendance
  (Inform Year Tutor)

- General concerns
  (Inform Year Tutor)

- Use of/access to technicians

- Guideline for assessment
  (criteria; double marking procedure; involvement of the external)

3.4   Monitoring progress

There is a range of ways by which students' progress can be monitored, including the traditional one-to-one supervisory meeting and records of progress, and a number of strategies for supervising large numbers of students.

Allocating supervisors

In a department with a variety of courses, or in a small department, it is very easy to arrive at an imbalance of supervisory workload, especially if the initial allocation is entirely based on staff or student preferences. As staff teaching loads increase, allocation of students to supervisors needs to be carefully controlled. Allocations often involve some balance between a number of considerations. Rank the following list in order of importance to you (and add to the list if you think that there is something missing).

Allocations may reflect:

- staff research/teaching interests;

- equity of workload among staff;

- student preferences;

- staff preferences;

- students and tutors are individually well-matched;

- weaker students are not allocated to the most inexperienced staff;

- particular courses are not unduly favoured in some way;

- larger supervisory load for staff with an aptitude for or interest in supervision.

Should all staff have the same level of supervision? Do all students need or receive the same level of supervision?

Supervisory meetings

One-to-one meetings between the student and tutor have been the traditional way in which student progress has been monitored but even here there is no fixed pattern. There are good pedagogic reasons, as well as the pressures introduced by increasing numbers of students, for reviewing the purpose of these meetings and how often they are necessary

"Our tutors were strict on deadlines, which was beneficial, because students are lazy!" — Ruth

Here is a checklist of issues associated with the organisation of the supervisory meetings. Which are the most important to you? Should others be added to the list?

- Are mutual expectations explicit and understood, perhaps as expressed in a Learning Agreement?

- Are the meetings appropriately scheduled? Do they need to be more frequent at the beginning? Are they so far apart so that if one is missed an undue amount of time passes without contact?

- Is account taken of other pressures on students? (e.g. work required in other subjects?)

- Are scheduled meetings which have been missed by either party followed up?

- Is there a mix of scheduled (formal) and as-needed (informal) meetings?

- Is each meeting prepared so that it is focused? Are the students asked to prepare for each meeting carefully?

- Is policy to be firm or flexible about interim deadlines?

- Is a record kept of each meeting to include, for example, date, time, agenda, goals met, difficulties, goals/actions agreed, other? (see below)

Records of progress and logbooks

Some geography departments use forms to record the progress of the supervision. Like Learning Agreements they are often a consequence of increasing student numbers, and a need to ensure some equity of support. Such forms often require signatures of both the tutor and the student for each entry.

An alternative (or complementary) method is the logbook, in which students write more extensively about their developing ideas. Logbooks are useful for getting students to reflect on their strengths, weaknesses and developing skills. In this sense they can be used to reinforce the value of dissertations as a learning experience. They typically also require both signatures for the relevant entries, notably the records of the supervision meetings and the handing over of drafts for comment.

Some staff feel that fixing obligations in writing like this can support the student with a clear framework to develop their organisational skills including goalsetting, prioritization, research skills, time management, and negotiating. Where departments particularly value the development of these skills, they may allocate part of the available marks to the logbook. One department allocates 10 per cent to it, taking the criteria as: completeness as a record and presentation; level of reflection; awareness of and the approach to goalsetting, prioritization, time management and negotiating.

> Is the record of progress or logbook useful merely as a means to an end? Should it be included in the mark allocation and, if so, how many marks should it carry? What about students who do not want or need close supervision?

Reading drafts

Not all departments allow students to submit drafts to tutors. Some argue that dissertation work is an individual study and that, apart perhaps from instructions about when it is to be done and how presented, it is up to the student to carry out the study entirely independently. In these departments, it is the view that the students are put to the test by being thrown back on their own resources, and that this pressure will bring out the best they have to offer.

Other departments are of the view that supervisory support is appropriate, not to compromise the student's independence, but to enhance her/his learning. Here, usual practice is that general advice is offered, and students' thinking prompted by questions. In doing this the supervisor might reassure the student that it is "safe" to get things "wrong" at this stage. The intellectual battle is being fought at the drafting stage, and the authenticity of the personal, even possibly original, end result will usually be at the price of some stumbling along the way. However, for many courses increasing student numbers preclude reading and re-reading drafted material, and some course teams stipulate the amount of drafted material a student can expect their supervisor to read (for example, two chapters).

Some of the key questions associated with reading drafts include: How important is it to read drafts? How can we provide the support needed without being overwhelmed with material to read? How much comment can we provide on drafts before we compromise the independent quality of the dissertation?

3.5 Supervising large numbers

The large numbers of students in HE will remain a live issue in the foreseeable future. Here are some strategies to maximize the return on time invested in the supervision process.

- *Group seminars*

  Use seminars as a forum for briefing on generic issues, perhaps, for example, in an introductory meeting. Other suitable points might be at the beginning of the final year (focusing on, for example, proposals, log books, schedules of meetings) and in the period leading up to the deadline (focusing on, for example, writing up, assessment criteria).

- *Prior identification of issues*

  Ensure that issues are clearly identified from the outset. It is useful to have a clear agenda available prior to each meeting, including for example, progress made towards targets, specific difficulties, or feedback on a previously-submitted draft. It may not be the best use of time to use the meeting to read a draft, unless it can be done rapidly.

- *Strict time limit/no interruption*

  Setting a time limit can be useful to focus discussion. If meetings have a strict time limit and no interruption is allowed considerable progress may be possible.

- *Office hours posted*

  Posting office hours can limit the fragmentation of work while working at your desk, but only if you adhere to the times. You might try to minimize responses to casual or "desperate" requests for a meeting.

- *Email*

  You might insist that primary contact is by an email message which could indicate the nature of the issue.

- *Student-led workshops*

  This is a good way of promoting mutual support. Although students will be dealing with different topics, they will be coping with a wide range of similar issues to do with organisational skills, availability of data or reading material, things found on the Internet, keeping notes and files, and tackling the drafting of individual sections. By sharing ideas the whole may be greater than the sum of the parts. Additionally, if the group gels well, individuals can report on their specific topic, and others invited to offer advice on the content.

- *Use a cassette recorder for feedback*

  If the student provides a blank tape, dictating your comments can be a quick and more personal way of feeding back, for example on drafts. More ground may be covered this way than in writing (Cryer & Kaikumba, 1987; Jenkins, 1993).

3.6   Assessing the dissertation

General issues associated with assessment are covered in another Guide in this series (Bradford & O'Connell, 1998). The issue here is the extent to which assessment of the dissertation differs from the assessment of other pieces of submitted work. Some course teams feel that the difference is great enough to necessitate a separate set of assessment criteria for the dissertation, and some student guides set out these criteria or provide a copy of the marking schedule given to tutors. Figure 2.2 shows an example. The issue, therefore, is whether the dissertation warrants its own assessment criteria, and if so how these differ from criteria for other parts of the course.

Assessing interim stages

Supervisors often only give general advice on drafts, but there are a number of ways in which assessment can be introduced before marking the final, submitted version of the dissertation. There are two main reasons for assessing at an interim stage. The first is to check on students' progress. Seminars which carry some of the marks for the dissertation can help to focus students' attention and can help to identify students who are poorly organised. The second reason is to assess elements not covered by the submission of a thesis such as, for example, the ability to communicate orally.

> How many marks should be allocated to interim stages in the preparation of a dissertation (seminars, record of progress, logbook, interim report, draft chapters)?

An alternative strategy is for the students to present an evaluation or assessment together with the draft, based on the published criteria. By responding to the evaluation rather than to the draft itself directly, the supervisor can be powerfully but legitimately supportive in a mature learning relationship.

Figure 2.2: *Assessing the dissertation*

DISSERTATION/EXTENDED ESSAY ASSESSMENT 1997   CONFIDENTIAL

Examiner's initials ....   Supervisor ☐   Second Examiner ☐

Student's Name   ....   **Degree   BA/BSc/BACH/BScJt/BSocSc**

Title

Criteria	Max mark	Comments; deviation from the mark scheme should be clearly indicated and reasons given	Mark
Research Problem and its Context Statement of problem; coherence and originality of aims	(5)		
Research context, including evaluation of previous research and provision of relevant local/ regional background	(15)		
Data Collection and Analysis Collection of data and appropriateness of evidence	(15)		
Analysis and interpretation of data/evidence	(15)		
Discussion of results and their implications	(15)		
Presentation Clarity of cogency of writing	(10)		
General presentation, including proof reading and the design, execution and use of maps, diagrams, tables and other illustrative material and their integration with the text	(10)		
Bibliography and citation, including relevance and accuracy	(5)		
Realisation of Aims i.e. in relation to both the stated aims and the potential of the topic	(10)		

Comments on the work as a whole: additional space is available overleaf.

Provisional mark ☐

Agreed mark ☐

ADDITIONAL INFORMATION

Further comments by supervisor

Any special circumstances or problems should be indicated here:

Comments by the Head of Department

Comments by the External Examiner

Double marking and the role of externals

> Table 2.3 is taken from one guide for tutors. How does the practice detailed here compare with practice in your department? Are all these rules necessary? Are they all the same as the rules for your course? Are they readily transferable? How can they be implemented?

Table 2.3: *An example of assessment regulations for the dissertation*

- All dissertations are double marked
- The dissertations tutor will supply a list of candidates which identifies their supervisor and second marker
- Where possible, second markers are chosen by the dissertations tutor to assess work which corresponds to their own research interest or expertise
- Each candidate submits their log book and two copies of their dissertation
- The dissertations tutor sets and circulates a date for returning candidates' marks to them
- The log book should be left in the secure room and used for reference only
- The supervisor should also provide the second marker with any relevant information concerning any special circumstances. Whilst dissertations cannot be marked 'sympathetically', this information helps set the work in context, and may be taken into account by the Examination Board
- Each marker should assess the work independently, awarding it a mark and generating a commentary of reasonable length on the work to justify their assessment
- The two markers should arrive at a joint mark.
- This joint mark should not simply be the mean average of the two independent marks, but should be reached in such a way that both assessors are satisfied that the dissertation's relative merits have been reflected
- If it is not possible to reach agreement, then a third marker, chosen by the supervisor, should be recruited to arbitrate
- Time should be allowed for the possibility of arbitration
- The third marker should assess the dissertation while blind to the assessments of the first two markers
- A further meeting should be arranged at which all three assessors agree upon a mark
- In the case of unresolvable difference, the work will be submitted to the external examiner for advice

References

4.1   References cited in the text

Bradford, M. & O'Connell, C. (1998) *Assessment in Geography* (Cheltenham: Geography Discipline Network, CGCHE)

Brown, S., Race, P., & Smith, B. (1996) *500 Tips on Assessment*, pp.56-57 (London: Kogan Page)

Burkill, S. & Burley, J. (1996) Directions: getting started on a geography dissertation, *Journal of Geography in Higher Education*, 20(3) pp.432- 433.

Cryer, P. & Kaikumba, N. (1987) Audio cassette tape as a means of giving feedback on written work, *Assessment and Evaluation in Higher Education*, 12, pp.148-153.

Gatrell, A.C. (1991) Teaching students to select topics for undergraduate dissertations in geography, *Journal of Geography in Higher Education*, 15 (1), pp.15-23.

Jenkins, A. (1993) Use of tape recorder to comment on student dissertation drafts, *Journal of Geography in Higher Education*, 17(2), pp.162.

4.2   Other sources

Boud, D. (1993) Three principles for good assessment practice, *New Academic*, 2(2), pp.4-5.

Gibbs, G. (1992) Down with essays!, *New Academic*, Spring 1992, pp.18-19.

Parsons, T. & Knight, P.G. (1995) *How to do your dissertation in geography and related disciplines* (London: Chapman & Hall).

Appendix 1
The importance of Health and Safety

Some courses include formal risk assessment training in preparation for the kind of fieldwork often associated with dissertation work. Some of the potential dangers or hazards in field research, which may be included in Fieldwork and Dissertation Guides, are:

- *What is a dangerous/hazardous location?*
  Guides might include a definition and examples: remote mountains or moorland, cliffs, caves, quarries, tunnels, potholes, spoilheaps, landfill, reedbeds, bogs, marshes, seashore, the inner city, 'no-go' housing estates, etc.

- *Preparation*
  Actions might include: discuss with tutor; leave itinerary (date and time of departure; method of travel to and at the location; itinerary; potential dangers/ hazards; ETA home)

- *Clothing and equipment*
  Suggestions might include warm clothes; cool clothes; protective clothing against the elements; helmet; goggles; food and drinks; map and compass; whistle; watch; torch; first aid kit; survival blanket.

- *Working alone in remote locations*
  Relevant cautionary suggestions may be appropriate

- *International distress signs*
  For mountains and other terrestrial locations; at sea.

- *Disease immunisation*
  Guides might include advice in connection with tetanus; risks associated with plants, animals, and freshwater; National Poisons Information Service; Labelling and registering samples.

An example of a risk assessment form, from the Cheltenham & Gloucester College of Higher Education, Department of Geography & Geology, is shown overleaf.

<table>
<tr><td>**DEPARTMENT OF GEOGRAPHY AND GEOLOGY**</td><td>**HEALTH AND SAFETY**
RISK ASSESSMENT FORM
(Management of Health and Safety
at Work Regulations, 1992)</td></tr>
</table>

1. MODULE TITLE AND CODE: GL211 - N. WALES MAPPING FIELDWEEK

2. LOCALITY: YR ARDDU (BEDDGELERT, SNOWDONIA) (SH 62 46; 62 47; 63 46; 63 47)

3. ASSESSOR: PHIL GRAVESTOCK

4. RISK ASSESSMENT DATE: APRIL 1996

5. DATE OF NEXT REVIEW: MARCH 1997

6. In the event of an accident or serious injury, contact the following:
- Ambulance service (if appropriate).
- Next of Kin.
- Department of Geography & Geology (01242 532971).
- College Health & Safety Officer, if serious injury involved (01242 532835; 0850 292515; 01242 517477).

7. Where should casualties be taken? Minfford Hospital (A&E), Minfford, Bangor (01248) 352308
Ysbyty Gwynedd Hospital (General), Penrhosgarnedd, Bangor (01248) 384384

8. What instructions/training is necessary for staff/students prior to visit?
- Use of personal protective equipment (PPE): YES - Hard hats, goggles, high visibility tabards.
- Working limits: YES - Mapping areas.
- Hammering: YES - Use of geological hammers, and goggles.
- Other: NO

9. Should any authorities/landowners be notified prior to visit? YES/NO

If yes, contact: _____

10. Is permission required before visit? YES/NO

If yes, contact: _____

11. Local weather conditions obtainable from: (0891) 500415 [long range 0891 449930]

12. Tidal information obtainable from: N/A

13. Local coastguard/mountain rescue: All accidents should be reported through the police (999).

Signature of assessor: *[signature]*

14. Are any of the following at risk?

- Staff   YES/~~NO~~   • Students   YES/~~NO~~   • Public   YES/~~NO~~

If YES, please comment on nature of risk: Risk of falling debris from cliff exposures. Uneven, rocky terrain, visibility reduced during bad weather conditions.

15. Does risk occur:

- Throughout visit to exposure/site: YES/~~NO~~ Cliff exposures - hard hats must be worn.
- Moving to and from exposures/sites: YES/~~NO~~ Uneven ground.
- During bad weather: YES/~~NO~~ Visibility reduced. Falling on slippery rocks. Risk of hypothermia.
- During adverse tides: ~~YES~~/NO _____
- From contractor's plant/operations: ~~YES~~/NO _____
- From passing traffic: ~~YES~~/NO _____
- Other instance (please specify): _____

_____

If YES, please comment on nature of risk. If more detail is required for any of these items please use a separate sheet.

16. Nature of hazard and extent of risk:

(e.g. steep slope = hazard; high risk = dangerous when wet; low risk = danger of person falling).

Hazard	Extent of Risk
1. CLIFF EXPOSURES	a. Danger of falling debris - moderate risk.
	b. Danger of person falling - low risk.
	c.
2. WALKING AROUND MAPPING AREA	a. Injury to bones and/or joints - moderate risk.
	b. Problem for asthmatics - low risk.
	c.
3. HAMMERING ROCK EXPOSURES	a. Accident during hammering - low risk.
	b.
	c.
4. BAD WEATHER	a. Accident during reduced visibility - moderate risk.
	b. Risk of hypothermia - moderate risk.
	c. Losing a member of the party - low risk.
5. STORMS	a. Danger of localised storm - low risk.
	b.
	c.
6.	a.
	b.
	c.

17. Personal protective equipment required:

- Hard hats: YES/~~NO~~   • Field boots: YES/~~NO~~   • Goggles: YES/~~NO~~
- Reflective waistcoats: YES/~~NO~~   • Gloves (disposable): ~~YES~~/NO   • Mask: ~~YES~~/NO   Survival Bag.
- Waterproofs: YES/~~NO~~   • Wellingtons/waders: ~~YES~~/NO   • Other (Specify) Whistles; First Aid kit

18. Specific risk and risk control measures (refer to numbers from Section 16):

Specific Risk	Risk Control Measures
1a. DANGER OF FALLING DEBRIS	Hard hats must be worn.
1b. DANGER OF PERSON FALLING	Students must not climb rock faces. Only look at rocks where the access is safe, both for climbing and descending.
2a. INJURY TO BONES AND/OR JOINTS	Take particular care when walking over rocky, uneven ground. There is no need to rush. The rocks in the area are particularly slippery during bad weather conditions (rain or fog). Try to avoid walking over rocky ground when possible.
2b. PROBLEM FOR ASTHMATICS	Asthmatics should take care when walking around the area (take inhaler).
3a. HAMMERING ACCIDENTS	Protective goggles must be worn when hammering, or when someone nearby is hammering.
4a. ACCIDENT DURING REDUCED VISIBILITY	High visibility tabards must be worn in bad weather. Care must be taken (see 2a).
4b. RISK OF HYPOTHERMIA	Suitable clothing must be worn (i.e. no jeans). Students are advised to carry hat/gloves/spare jumper/flask of hot fluid/emergency rations, e.g. Mars bar. Each group must carry a survival bag.
4c. LOSING A MEMBER OF THE PARTY	High visibility tabards must be worn in bad weather conditions. Always know the location of other members in the individual mapping groups.
5a. DANGER OF LOCALISED STORMS	Leave the hillside immediately if hair starts standing on end (static in the air).

CHELTENHAM
&
GLOUCESTER
COLLEGE OF HIGHER EDUCATION